George Washington Carver

A Buddy Book
by
Rebecca Gómez

ABDO
Publishing Company

VISIT US AT

www.abdopub.com

Published by Buddy Books, an imprint of ABDO Publishing Company, 4940 Viking Drive, Suite 622, Edina, Minnesota 55435.

Printed in the United States.

Edited by: Christy DeVillier
Contributing Editors: Matt Ray, Michael P. Goecke
Image Research: Deborah Coldiron
Graphic Design: Jane Halbert
Cover Photograph: Iowa State University Library (Special Collections)
Interior Photographs: Henry Ford Museum, Iowa State University Library (Special Collections), Library of Congress, Photospin, National Portrait Gallery, Tuskegee Institute

Library of Congress Cataloging-in-Publication Data

Gómez, Rebecca.
George Washington Carver / Rebecca Gómez.
p. cm. — (First biographies. Set III.)
Summary: An introduction to the life of George Washington Carver, who was born a slave in Missouri and went on to become a college professor known for his accomplishments in the field of agriculture.
Includes bibliographical references and index.
ISBN 1-57765-944-9
1. Carver, George Washington, 1864?-1943—Juvenile literature. 2. African American agriculturists—Biography—Juvenile literature. 3. Agriculturists—United States—Biography—Juvenile literature. [1. Carver, George Washington, 1864?-1943. 2. Agriculturists. 3. Scientists. 4. African Americans—Biography.] I. Title.

S417.C3 G66 2003
630'.92—dc21
[B]

2002074676

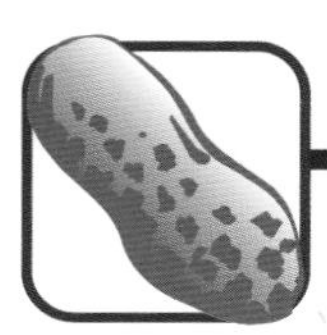

Table Of Contents

Who Is George Washington Carver? 4

George's Family 6

The Plant Doctor................8

Going To School 10

College 15

Tuskegee 20

Carver And Peanuts................24

World Famous 26

Important Dates................30

Important Words................31

Web Sites 31

Index 32

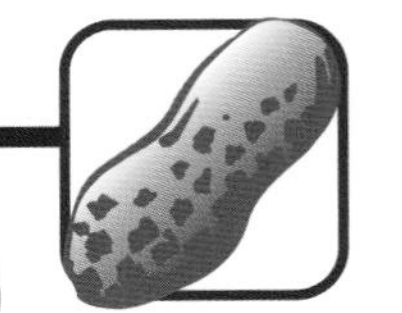

Who Is George Washington Carver?

George Washington Carver was a famous scientist. He invented many new things from peanuts, soybeans, and sweet potatoes. Some of these things are plastic, shoe polish, and peanut butter.

George Washington Carver

George Washington Carver was a teacher, too. He taught people how to be better farmers.

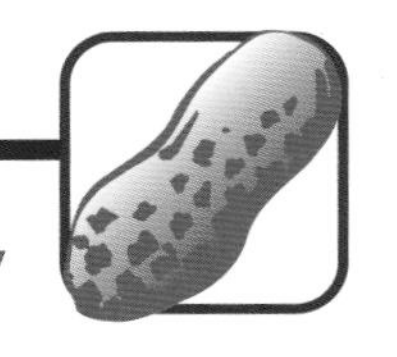

George's Family

George Carver was born near Diamond Grove, Missouri. He was born around 1864. No one is sure of his birthday.

George and his family were slaves. Susan and Moses Carver owned them. In 1865, slavery ended and George's family was free.

Young George lived in a log house.

George never knew his father. His mother was kidnapped when he was a baby. So, George and his brother Jim lived with Susan and Moses Carver. The Carvers treated them like their own children.

The Plant Doctor

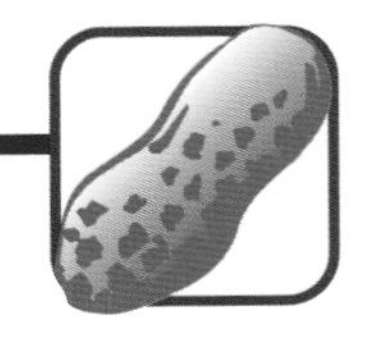

Young George liked to walk in the woods. He enjoyed studying the plants and insects he found.

George had a talent for growing things. He grew a special garden for himself. George learned which plants liked the sun. And he learned which plants favored shade. The people of Diamond Grove began calling George the plant doctor.

George Carver (left) later became a "plant teacher."

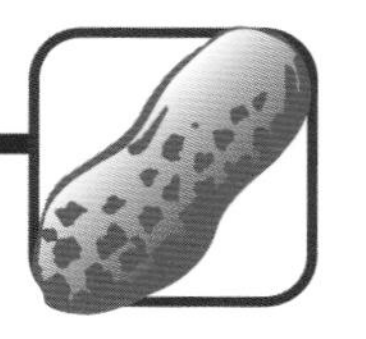

Going To School

George wanted to go to school. But the only school in Diamond Grove was for white students. Back then, African-Americans like George went to African-American schools.

In 1877, George left Diamond Grove. He went to the African-American school in Neosho, Missouri. It was eight miles away from Diamond Grove.

George Carver with his classmates at Iowa Agricultural College.

George found some nice people to live with in Neosho. Mariah and Andy Watkins cared for George. He helped Mariah around the house.

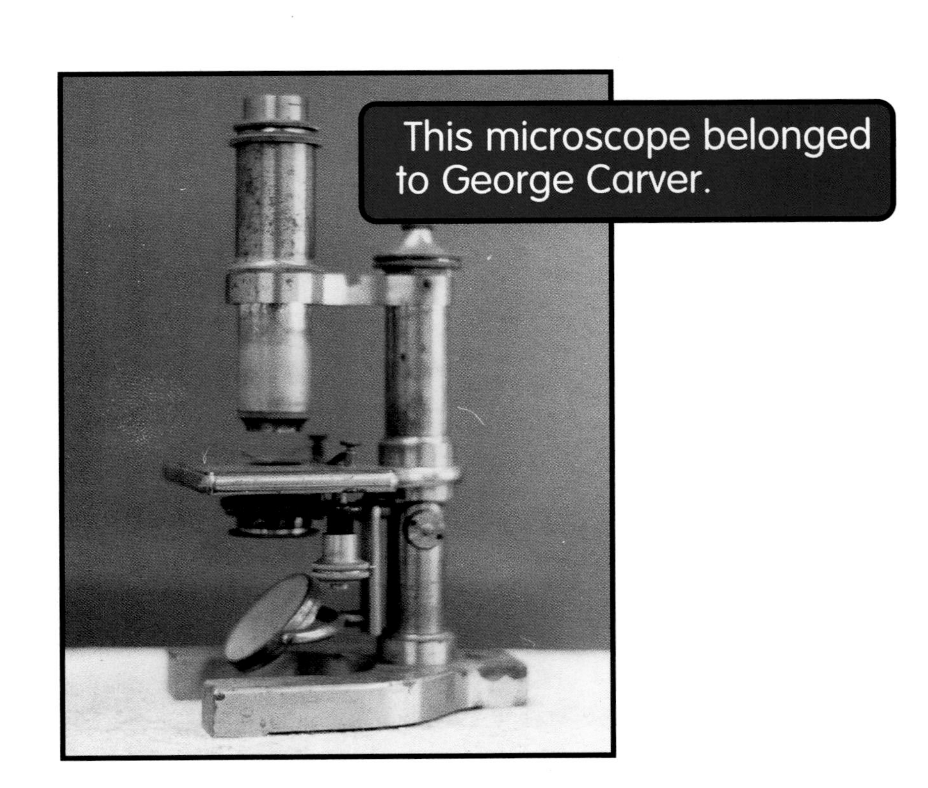

This microscope belonged to George Carver.

George learned all he could in Neosho. Then, he moved to Kansas. George went to many different schools in Kansas. George worked at different jobs, too. This was not an easy way of life. But George wanted to keep learning new things.

In Olathe, Kansas, there was another person named George Carver. People were mixing up George with the other George Carver. So, George gave himself a middle name. This is how George Carver became George Washington Carver.

George Carver gave himself the middle name Washington.

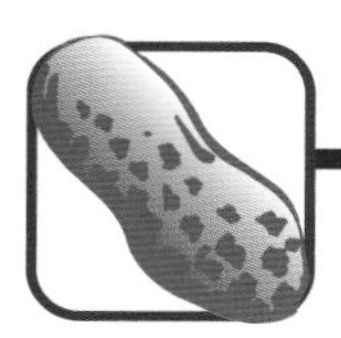

College

Around 1884, George Washington Carver finished high school. He wanted to go to college next. George got a letter from Highland College that said he could go there.

George showed up for school. But Highland College told him to leave. Highland College did not want any African-American students.

After high school, George bought some land and grew wheat.

In 1890, George tried going to college again. He went to Simpson College in Iowa. There, George studied art. George's teacher discovered that George knew a lot about plants. Her name was Etta Budd.

Etta Budd (bottom) with her family.

Etta Budd told George he should go to Iowa Agricultural College. There, he could study plants. George Washington Carver was the first African-American to go to Iowa Agricultural College. He studied plants and agriculture. Later, George became the first African-American teacher there.

George Washington Carver became a talented plant scientist.

Tuskegee

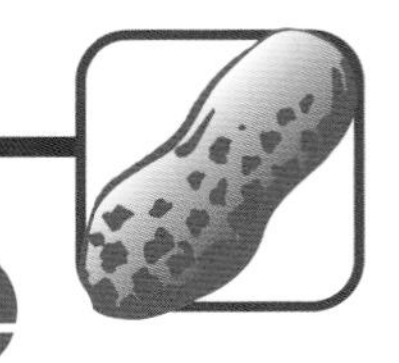

In 1896, George Washington Carver began teaching at the Tuskegee Institute. This was a college for African-Americans in Tuskegee, Alabama. Booker T. Washington started this school in 1881.

George Washington Carver (top left) at Tuskegee Institute.

Stamp of Booker T. Washington

Carver was the head of agricultural studies at Tuskegee Institute. He taught his students about plants and soil. Carver brought his students outside to learn about nature. He cared about his students and they learned a lot.

Carver studied plants in his free time.

Carver showed farmers new crops to grow.

Carver helped nearby farmers, too. He taught them how to take care of their soil. Farmers learned to rotate crops. Carver showed them new crops to grow. Southern farmers began planting black-eyed peas, sweet potatoes, and soybeans.

Carver And Peanuts

At one time, people mostly used peanuts to feed farm animals. Carver invented over 300 more uses for peanuts. He made medicines, soaps, and many foods from peanuts. Carver's peanut oil helped people suffering from an illness called polio.

George Washington Carver made peanuts famous.

World Famous

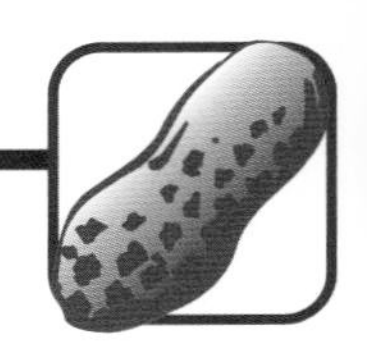

George Washington Carver's discoveries made him famous. The Royal Academy of Arts in England made Carver a member. In 1939, President Franklin D. Roosevelt gave Carver a medal. President Roosevelt thanked Carver for all his hard work.

George Washington Carver (left) with Henry Ford, another famous inventor.

Carver enjoyed sharing what he knew with people. He could have made a lot of money from his inventions. But being rich was not Carver's goal. Helping people made him happy.

On January 5, 1943, George Washington Carver died. People today still use many of his inventions.

Some of Carver's other famous inventions are:

- ✓ buttermilk
- ✓ chili sauce
- ✓ ink
- ✓ instant coffee
- ✓ mayonnaise
- ✓ metal polish
- ✓ shaving cream

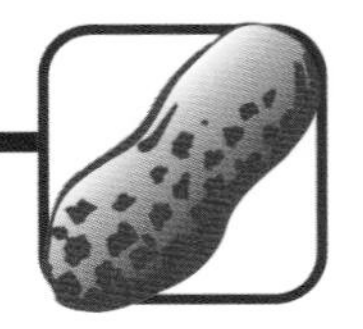

Important Dates

1864 George Carver was born around this time.

1865 Slavery ends, freeing George and his family.

1877 George goes to Neosho, Missouri, to begin school.

1890 George becomes the first African-American student at Simpson College in Iowa.

1894 Carver becomes the first African-American to graduate from Iowa Agricultural College.

1896 Booker T. Washington invites Carver to teach at Tuskegee Institute.

1938 The George Washington Carver Museum opens at Tuskegee Institute.

1939 President Franklin D. Roosevelt gives Carver a medal for his discoveries in science.

January 5, 1943 George Washington Carver dies.

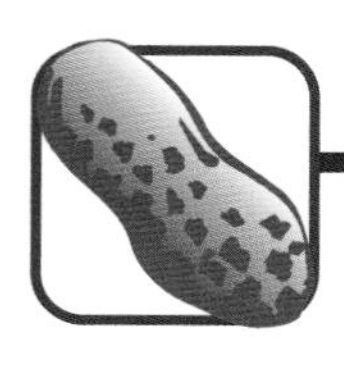

Important Words

African-American an American whose early family members came from Africa.

agriculture the study of farming.

invent to make something new. These new things are called inventions.

kidnap to carry off someone against their will.

rotate crops to plant a different crop in the same piece of land each year. Rotating crops is good for the soil.

scientist a person trained in science.

slave a person who can be bought and sold.

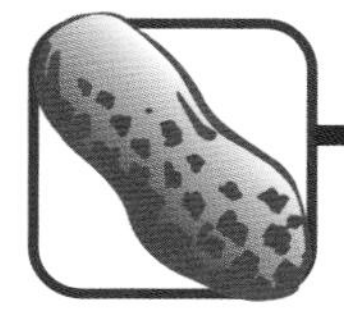

Web Sites

The Legacy Of George Washington Carver
www.lib.iastate.edu/spcl/gwc/bio.html
Learn more about this famous inventor from Iowa State University, which used to be Iowa Agricultural College.

George Washington Carver
http://inventors.about.com/library/weekly/aa041897.htm
Facts about George Washington Carver and other famous inventors are featured here.

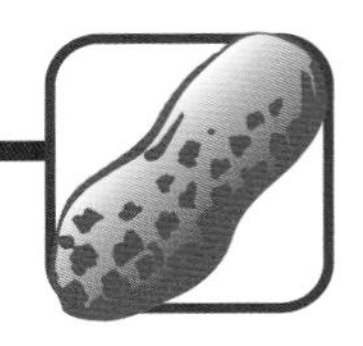

Index

African-American **10, 11, 15, 18, 20, 30**

agriculture **18, 30**

black-eyed peas **23**

Budd, Etta **17, 18**

Carver, Jim **7**

Carver, Moses **6, 7**

Carver, Susan **6, 7**

Diamond Grove, Missouri **6, 8, 10, 11**

Ford, Henry **27**

Highland College **15**

Iowa **17, 30**

Iowa Agricultural College **11, 18, 30**

Kansas **13, 14**

medicines **24**

Neosho, Missouri **11, 12, 13, 30**

Olathe, Kansas **14**

peanut butter **4**

peanuts **4, 24, 25**

plastic **4**

polio **24**

Roosevelt, President Franklin D. **26, 30**

rotate crops **23**

Royal Academy of Arts **26**

scientist **4, 19**

shoe polish **4**

Simpson College **17, 30**

slaves **6**

soaps **24**

soybeans **4, 23**

sweet potatoes **4, 23**

Tuskegee Institute **20, 22, 30**

Washington, Booker T. **20, 21, 30**

Watkins, Andy **12**

Watkins, Mariah **12**